ISBN 978-1-105-59713-8

90000

9 781105 597138

The Cholesterol Chronicles:

Uncovering the Truth-What Your M.D. Doesn't Know About Statin Drugs and Neuropathy

About the Author:

Dr. Joe DiDuro is dedicated to helping people live a more fulfilling life without pain and disease. His passion is to help people by teaching them how they can create peace in their own lives.

Dr. DiDuro earned his Bachelor of Arts degree from the State University of New York at Buffalo with a double major in Chemistry and English. He received his Doctor of Chiropractic Degree from Palmer College of Chiropractor in Davenport, Iowa.

Dr. DiDuro originally practiced in his hometown in Upstate New York where he worked for 8 years. His brother, also a chiropractor, took over this practice when Dr. DiDuro, who has a dual citizenship, moved his whole family to live and practice in Vicenza, Italy for 9 years.

While in Italy, he served as a director of a multi-disciplinary clinic. He cared for people in a country where only 200 chiropractors existed. He also loved experiencing the culture and people of Europe.

Dr. DiDuro pursued additional studies in Europe in the Neuropathophysiology department of the University of Verona Hospital and is, currently, the only doctor of chiropractic to present research at the Society for Italian Neurologist.

Dr. DiDuro also completed further training in Amsterdam for the American Chiropractic Associations' American Board of Chiropractic Neurology and has Diplomate Status in Chiropractic Neurology. While a Diplomate in Neurology is adapted to treating a myriad of difficult cases like chronic pain, his current area of clinical interest is Peripheral Neuropathy, Balance Disorders and Restless Leg Syndrome.

Dr. DiDuro then returned to the US to attend University and completed an NIH funded program for a Master's Degree in Clinical Research from Palmer Center for Chiropractic Research. He is one of a handful of clinicians that is qualified in the field of Chiropractic Research.

Dr. DiDuro, Chiropractic Neurologist, Clinical Researcher has worked with patients suffering from Peripheral Neuropathy for the last 10 years. He has published and presented papers, books and articles relating to chiropractic and health concerns. He has presented his research all over the world, including the U.S., Canada and Europe. These publications document the discoveries that have helped thousands of people get their lives back.

Currently, Dr. DiDuro is a member of the prominent scientific society, the Peripheral Nerve Society. He is the Founder and President of the Neuropathy Treatment Centers of America. Dr. DiDuro, in his years of clinical experience working with peripheral neuropathy patients, has come to understand the very low quality of life they experience – that the pain of neuropathy touches every aspect of a person's existence.

About the Neuro TCA

The Neuropathy Treatment Centers of America (NeuroTCA) is a 501(c)(3), non-profit organization whose main goal is to help create a neuropathy free world.

The NeuroTCA, with its connections to state-of-the-art technologies, collaboration with world renowned scientists and a dedicated group of active, certified neuropathy professionals, offers hope where suffering patients often see none.

Through education and increasing awareness among physicians, researchers, institutions and the general public, the Neuro TCA strives to inform the world that there are therapies, protocols and educational means to overcome the suffering that peripheral neuropathy pain often brings.

OUR VISION

Neuropathy Treatment Centers of America (NeuroTCA) will be a place of patient empowerment and advocacy ensuring that neuropathy patients have a place to share their stories of life-changing care and actively make a difference to the positive, progressive actions in developing treatments.

OUR GOALS

The positive forward thinking care offered to patients connected through our site and with the diligent work of our dedicated staff, we educate patients from all over the world on the perhaps little known, clinically significant breakthroughs that offer patients better quality of life.

While some groups focus on coping with pain and suffering caused by peripheral neuropathy, the NeuroTCA makes forward strides in participating in and funding research in the treatment and recovery for peripheral neuropathy, but with a goal of prevention. By providing patients choices for care around the globe and spreading information on the breakthroughs in technologies and science, we strongly believe that we can create a neuropathy free world.

Visit us online @

www.neurotca.com

A Message from Dr. DiDuro

I see patient after patient in my practice who is struggling with the consequences of high cholesterol treatments. Often, they are unaware that there are ways to lessen their unnecessary suffering. Now, I could tell you that it is as easy as taking a pill, but that would be a lie. However, the cure is natural and is much better for you than taking pills.

So, if you have been told that you have high cholesterol and that you should be taking a statin drug, this is the booklet for you!

Introduction - What inspired this book:

I am a chiropractic neurologist, clinical researcher, founder and president of the Neuropathy Treatment Centers of America. I have devoted much of my professional career to working with patients suffering from peripheral neuropathy, helping them regain their lives after experiencing the devastating side-effect of statin drug use.

Patient after patient has entered my office, suffering unnecessarily from the consequences of statin drugs prescribed by their physician. Often, their symptoms become debilitating. Neuropathy, muscle pain, memory loss, dizziness, and even depression are some of the most common symptoms experienced by patients using statin drugs. Many of my patients have suffered with statin-induced symptoms for months, if not years. Yet, they never made the connection between the two.

During my years of clinical experience, I have gained empathy and understanding for my patients. I have witnessed the low quality of life created by their condition. Whether neuropathy has been caused by statin drug use or some other source, it touches every aspect of their lives. I have watched them become victims of medical terrorism. It is my goal to provide them with the ammunition they need to fight back for their right to be healthy.

I first became interested in the relationship between neuropathy and statin drug use in the year 2000. I was practicing in Italy. My mother, 78 years old and a diabetic, experienced an incident that perfectly highlights the dangers of statin drugs. She had traveled to San Francisco for a vacation. She had a wonderful time and especially enjoyed riding the trolley cars up and down the city streets.

While boarding the flight home, she seemed in good spirits and in good health. The flight itself was uneventful. Yet, when the plane landed, she discovered that she was unable to get out of her seat. Without any apparent explanation, she had suddenly lost her ability to walk. She could not move at all.

For the next three months, she could not get out of bed or walk without the use of a cane. She experienced a great deal of pain in her legs. My brother, a chiropractor, treated her but was unable to help her. I urged her to get an MRI. After several weeks of me coaxing, she finally went. When the test results came back, she was informed that she simply had a "bad back."

The story could have ended there, with her learning to live with a bad back and an inability to walk, but it didn't. She had other tests, including an electromyogram (EMG), a test that measures electrical activity in the muscles to determine nerve function. She had needles stuck in her. She had routine neurological exams, nerve conduction tests, along with a variety of other tests. They all came back normal. When all of the tests were done and no source for her pain and numbness could be identified, the doctors were stumped. They told her, "Mickey, we see no reason why you can't walk. You aren't sick enough to be this sick." It was a frightening and frustrating experience and my mother was devastated.

I had seen this same scenario before, at my practice in Italy. In that practice, we had a full rehabilitation clinic; four university-based neurologists and EMG machines in a country where the chiropractic profession was not recognized by the mainstream medical community. We performed the very same tests on our patients that doctors were now performing on my mother.

In our patients, we had come up with the same responses as my mother's physicians. However, these standard tests, performed on thousands of patients in similar situations, only address large nerve fiber issues. They do not take potential small nerve fiber issues into account.

After weeks of testing provided no answer for my mother, I finally asked my mother if she happened to be taking statin drugs. As it turned out, she had been taking a common statin for only five or six months. Knowing that there was research that was now starting to show a link between statin drugs and nerve damage, I told her, "You better get off that thing."

She stopped taking the statins and, before long, was able to move again. She underwent chiropractic care at my brother's office. Within just a few weeks, she was able to walk again, but the damage had already been done. It was severe, after only a few months of taking the drugs. To be fair, she did have other risk factors. She was a diabetic and that alone carries with it a greater risk of neuropathy and it was likely that she already had some nerve damage. Once she started taking the statin drugs, they hit the weakest nerves first and accelerated the problem.

After my mother's incident, I became even more interested in the link between statins and peripheral neuropathy. I had already seen so many diabetic patients with nerve damage in my office. I learned that, with many of them, their neuropathy worsened once they began taking the statins. This only further increased my curiosity. I began studying the research and even helped to collect data for a major research project in California.

I have internationally published and presented several papers on my experiences and my research. These papers document the discoveries that have helped thousands get their lives back by finding relief from their neuropathy pain. While neuropathy impacts millions of lives each year, my research still falls through the cracks. The promotion of statin drugs has drowned out my efforts. The glossy magazine and television ads tell us that statin drugs are the answer "when diet and exercise are not enough." According to one drug advertisement, the drugs carry with them some "little-known side effects and few known risks." Is this true or are the drug companies minimizing the potential risks?

Everyone has something to say about cholesterol. However, this talk is usually used to sell something-a drug, a cure, a supplement. The list goes on. This book is not aimed at selling anything. It aims to set the record straight while shedding some light on research that has been done on statin drugs and the dietary therapies that can help lower cholesterol naturally, without reliance on drugs. For many readers, the truth is going to be an eye opener, and quite possibly life-changing. As the saying goes, "the truth will set you free." It is my hope that, by knowing the truth, patients will simply find the drug-free solution that helps them feel better.

This book is a culmination of years of research. This research examines the relationship between statin drugs and neuropathy. It also looks at how common statin drug use has become, even though it is often not necessary. The majority of patients can easily lower their cholesterol, on their own, through diet and exercise.

My book is written for anyone who struggles with neuropathy or any other adverse effects that are often experienced during drug therapy for high cholesterol. Although the cure this book offers is not as easy as taking a pill, it is a solution that can be life-changing for anyone who wants to improve their life, end their pain and start feeling better.

In my office, I began offering testing that included a full neurological workup and an exam that offers help to people who suffer from small fiber neuropathies. For many of them, the therapy I offered gave them the specific neurorehabilitation they needed to take back their lives. They were able to ditch their pricey statin prescriptions and keep their cholesterol levels low. More importantly, these patients regained normal sensations in their legs and feet. They were finally able to move again. This book will show you how you can do it too.

Terminology

This book might contain a few words that are new to you. For easy reference, a list of these terms has been provided for you here.

Hypercholesterolemia / Hyperlipidemia:

The medical term for the presence of high levels of cholesterol in the blood.

Statin Drugs:

Medication that lower blood cholesterol levels by inhibiting cholesterol-making liver enzymes.

LDL:

The acronym for Low Density Lipoprotein. LDL is also commonly referred to as "bad cholesterol." This is the number that folks often aim to keep low. HDL, or High Density Lipoprotein, is sometimes called "good cholesterol.

Triglycerides:

Fatty substances that are present in the blood.

Peripheral Neuropathy:

A condition that results from damage to the peripheral nerves. Symptoms include weakness, numbness and pain to the hands and feet.

Polyneuropathy:

A neurological disorder that occurs when many peripheral nerves throughout the body malfunction simultaneously.

Psyllium Husks:

A dietary fiber supplement that aids in digestion and lowers cholesterol. Psyllium, sometimes referred to as "ispaghua," are the seeds from the Plantago plant. As a dietary supplement, psyllium husks are sold over-the-counter in a variety of forms. Seeds are sometimes ground and used in gluten-free baked goods.

Myelin:

A fatty substance made of lipids and lipoproteins, which protects the nerves. It is essential for nerve function.

EMG:

Abbreviation for electromyogram, a test that measures electrical activity in the muscles to determine nerve function.

Plantar Fasciitis:

Also diagnosed along with heel spurs. Plantar fasciitis is a painful foot condition caused when the nerves connecting the heel to the toes become inflamed. Many people are misdiagnosed with plantar fasciitis when, in fact, they are suffering from neuropathy.

Nerves:

A network of fibers within the body that relay messages between the body's organs and the brain and spinal cord.

Commonly Prescribed Statin Drugs:

Commonly prescribed brand-name statin drugs include Lipitor, Zocor and Crestor, as well as many others. Generic statin drugs are manufactured under these brand-names. Some examples include:

Atorvastatin is marketed under the brand name Lipitor. Fluvastatin is marketed umder the brand name Lescol. Lovastatin is marketed under the brand names Mevacor and Altocor. Pravastatin is marketed under the brand names Pravachol, Selektine, and Lipostat. Rosuvastatin is marketed under the brand names Zocor and Lipex. Cerivastatin is marketed under the names Lipbay and Baycol.

Chapter 1:

Statins and Neuropathy

Research indicates that the most common side effects of statin use are:

- Muscle pain
- Muscle weakness
- Fatigue
- Neuropathy
- Cognitive reduction
- Peripheral neuropathy

Suspected side effects are:

- Sleep disorders
- Erectile dysfunction
- Psychiatric, mood or irritability issues

The U.S. Food and Drug Administration deemed that the drugs are safe when used for their intended purpose - lowering cholesterol.

However, as the use of statin drug use has increased steadily over the years, due to the increasing numbers of individuals dealing with high cholesterol, we are now gaining better insight into their effects on the human body. As a result, statin-induced neuropathy is becoming more commonly mentioned in scientific literature.

A typical prescription for a statin drug comes with a warning label that reads: "Unexplained muscle pain and weakness could be a sign of a rare but serious side effect and should be reported to your doctor right away." On its website, the FDA warns readers that common side effects of statins include gas, dizziness, constipation, headache, diarrhea, and upset stomach [Reference 1 http://www.fda.gov/Drugs/DrugSafety/PostmarketDrugSafetyInformationforPatientsandProviders/ucm204882.htm#AdditionalInformationforPatients]. The regulatory agency also warns readers about statin-related muscle pain, tenderness or weakness and rhabdomyolysis, a condition that could lead to kidney damage and death.

Risks are greater for patients who drink more than a quart of grapefruit juice a day or for folks that take other medicines for heart or stomach conditions, antibiotics, antifungals or a variety of other diseases. People should heed the warnings because, in extreme cases, the side effects of statin use can even lead to death.

Chiropractors are on the front lines of dealing with these issues. Chronic pain and pain that is not relieved by medication often leaves patients exasperated, driving them to seek out alternative cures like chiropractic care, in hopes of finding relief. Unfortunately, some chiropractors may, unknowingly, fail their patients by focusing on spinal malfunction rather than toxic neuropathy. The seriousness of this condition, paired with the growing number of patients dealing with peripheral neuropathy has created a serious need for chiropractors to understand this condition so that they can better help their patients.

Symptoms of neuropathy might include, but are not limited to: pain, numbness, tingling or prickling sensations, burning pain (especially at night), sensitivity to touch, restless leg syndrome, balance disorders and difficulty walking. If left untreated, neuropathy can eventually lead to deterioration of the muscles and paralysis. Being a chiropractor neurologist, I found that statin drug patients would complain of these symptoms, but were unaware that what they were describing was neuropathy. Even more disturbing was the fact that they didn’t realize that this pain, for many individuals, was caused by statin drugs prescribed by their doctor.

How Do Statins Work?

Cholesterol is produced naturally, by the liver. The human body is designed to handle natural cholesterol production but sometimes the body produces too much, putting stress on the body. The statin drug's primary job is to block cholesterol-producing enzymes. By blocking cholesterol production, statins help remove more cholesterol from the bloodstream, lowering the risk of high-cholesterol linked diseases like cardiac disease.

Researchers believe that, in interfering with cholesterol synthesis, the drugs also alters myelin and nerve function. Some researchers estimate that 1 in 10 statin users will develop at least mild neuropathy with symptoms including muscle pain, weakness and or fatigue. These patients may have difficulty rising from a chair or getting out of bed. They may even experience shortness of breath, causing difficulty when walking.

In our practice in southern Arizona, we took a closer look at the statistics of statin drug users who suffer from neuropathy. We analyzed a group of patients who came into the office during a three month period. We screened 246 patients over the age of 50 for peripheral neuropathy. Of the patients that were screened for peripheral neuropathy, 59 percent were female and:

- 30.9 percent were 50 to 63 year olds
- 46.2 percent were 64 to 74 years of age
- 22.9 percent were 75 years or older

As a whole, and astonishing 48.4 percent of the group demonstrated positive tests for some form of peripheral neuropathy. Of that group:

- 19 percent of those individuals were diabetic
- 36.6 percent were using statin drugs

This completely showed light on the epidemic I was seeing in my office and I wondered if this was true across the country. I wanted to explain what I found so that Chiropractors and medical practitioners could be aware of the damaging effects that statin drugs can have on their patients' health. This is especially true when considering percentage of people taking statin drugs in the general public, and in those entering the chiropractic clinics. Proper diagnostics and treatment of drug-induced neuropathy is essential to increasing the quality of life in these individuals. By understanding neuropathy, chiropractors can help educate patients about their condition. They can ensure that their patients receive the treatment the need and deserve.

Chapter 2

How do Statin Drugs Affect the Body?

Small countries like Denmark and Iceland are great for conducting studies on medical conditions. Because the population is all genetically similar, unlike the vast melting pot in the United States, it is easier to conduct a controlled study and get reliable results.

One Danish study found that statin users were four times more likely to develop peripheral neuropathy than non-statin users. It was also found that patients who took statins and already had some existing peripheral neuropathy had a risk that was 16 times greater than non-users. While long term statin use is considered to be two years or longer, nerve damage can happen within just a few days of starting the drug.

Overall, the study concluded that “long-term exposure to statins may substantially increase the risk of polyneuropathy.” Despite these findings, however, the study declared that “the positive benefits of statins, particularly on reducing the risk of heart disease, far outweigh the potential risks of developing neuropathy.” University of Southern Denmark researcher, David Gaist, also stated that “these findings shouldn’t affect the doctor or patient decisions to start using statins. But if people who take statins develop neuropathy symptoms, they should talk with their doctor, who may reconsider the use of statins.” [Ref 7 http://www.neurology.org/content/58/9/1333.abstract

http://www.medscape.com/viewarticle/433446]

Dr. Beatrice Golomb, however, studied the negative effects of statin drug use to determine if the benefits truly outweighed the side effects that patients often experience when taking these drugs. The study used patient feedback to determine the results. https://www.statineffects.com/info/

I was practicing in Iowa at the time, working on my Master’s Degree in Clinical Research from Palmer College of Chiropractic Research. This was a National Institutes of Health funded program that turned clinicians into researchers. I had been waiting for this chance for most of my career. I had read about Dr. Golomb's research while studying statin side effects during my years in Italy. Now, being back in the states, I was able to participate in the study. It was very satisfying to help a fellow researcher gather data that might shed light on the epidemic.

During my participation in the study, patients who came into my practice were asked to fill out a questionnaire from Dr. Golomb's study. I made the questionnaire available both in my office and online. The study offered information and addressed some of the unusual symptoms that can often occur with statin use. In my practice, I found that, not only were people having muscle pains and memory loss, but they were more likely to have peripheral neuropathy or pain in their feet.

While not a member of the actual study, one of my patients, James G., a 77-year-old man, was one of the many patients that reported this problem. He once approached me after a lecture and said, "I've got to talk to you. I think I've got a problem. I've got plantar fasciitis."

Plantar fasciitis, also commonly associated with heel spurs, is one of the most common causes of heel pain. It is caused when the plantar fascia – a tissue that connects the heel bone to the toes – becomes inflamed, resulting in a stabbing pain in the bottom of the foot. People usually feel this pain when they get out of bed in the morning or after sitting for long periods of time. It can come and go and it is most common in people who are overweight, marathon runners, and people who wear shoes that do not have adequate support.

I asked James if he was a marathon runner, to which he replied that he was not. I could clearly see that he was not overweight. And, although the mainstream medical community would tell this man that he had plantar fasciitis and treat him accordingly, despite the fact that he does not fall into any of the risk groups, I suspected that his foot pain was not caused by plantar fasciitis, but by peripheral neuropathy.

Neuropathy is a condition that is under-diagnosed and under-reported. Not only are patients often unaware that they are taking a drug that will give them problems, when they go to the doctor, they are given the wrong diagnosis. Imagine the impact that this wrong diagnosis will have on their lives.

Neuropathy is the main focus of my practice but the problems with statins can extend far beyond neuropathy. Research conducted by Dr. Beatrice Golomb M.D, Ph.D. found that statins can affect the brain's performance and cause other problems, including memory loss, confusion and muscle pain.

An even bigger problem, however, comes with older patients. In many cases, they have been taking statins longer and have other health issues that are complicated by statin use. I've seen many examples in my own practice.

According to a study conducted by Dr. Golomb in May of 2005, "The elderly differ from younger people in the relation of cholesterol to heart disease and mortality. Clinical trial evidence supports epidemiological findings in showing that the risk of high cholesterol weakens in relationship to heart disease with age and loses its relation to mortality." REF: Golomb BA. Impact of statin adverse events in the elderly. Expert Opin Drug Saf. 2005;4(3):389–397. What she is saying is that cholesterol is not seen as a risk factor for heart attacks and strokes in the elderly.

But, she continues this study also determined that "the elderly may be more vulnerable to known AEs (adverse effects) and evidence provides cause for concern that new risks may supervene, including cancer, neurodegenerative disease and heart failure." Essentially, this study indicates that, in elderly patients, statin drug use could actually cause the very same health issues that they are intended to prevent. What's more, elderly patients seem to be even more prone to the serious side effects of statin drugs.

Not only were the side effects in all patients found to be serious with statin drug use, Golomb also found that the bulk of positive press stories about statins focused on studies that were funded by drug companies. The fact that many patients "lost thinking ability so rapidly that, within the course of a couple of months, they went from being heads of major divisions of companies to being unable to balance a checkbook and being fired from their companies" was lost in a flurry of media attention that touted the benefits of the medication.

Despite the positive media portrayal of statin drug use, Dr. Golomb's research was eventually featured on ABC News "Good Morning America." The segment also included an interview with GMA medical correspondent, Dr. Timothy Johnson who shared his thoughts about the study. Johnson said that drug manufacturers recognize that about 2 percent of patients taking statins would develop some memory loss or confusion, which mimicked Alzheimer's disease. Johnson said that, since the brain is "full of cholesterol," that it is theoretical that a drug that lowers cholesterol could potentially impact the brain tissue.

Another study conducted by Dr. Golomb, published Tuesday, Jan. 27, 2009 in *Health and Medicine* [Reference http://www.ncbi.nlm.nih.gov/pmc/articles/PMC2849981/], examined more closely the actual relationship between statin drugs and the body.

> *A paper co-authored by Beatrice Golomb, M.D., Ph.D, associate professor of medicine at the University of California, San Diego School of Medicine and director of UC San Diego's Statin Study group cites nearly 900 studies on the adverse effects of HMG-CoA reductase inhibitors (statins), a class of drugs widely used to treat high cholesterol. The result is a review paper, currently published in the on-line edition of 'American Journal of Cardiovascular Drugs', that provides the most complete picture to date of reported side effects of statins, showing the state of evidence for each. The paper*

also helps explain why certain individuals have increased risk for such adverse effects.

"Muscle problems are the best known of statin drugs adverse side effects," said Golomb, "but cognitive problems and peripheral neuropathy, or pain or numbness in the extremities like fingers and toes, are also widely reported." A spectrum of other problems, ranging from blood glucose elevations to tendon problems can also occur as side effects from statins.

The paper cites clear evidence that higher statin doses are more powerful statins- those with a stronger ability to lower cholesterol – as well as certain genetic conditions, are linked to greater risk of developing side effects.

"Physician awareness of such side effects is reportedly low," Golomb said. "Being vigilant for adverse effects in their patients is necessary in order for doctors to provide informed treatment decisions and improved patient care."

The paper also summarizes powerful evidence that statin-induced injury to the function of the body's energy-producing cells, called mitochondria, underlies many of the adverse effects that occur to patients taking statin drugs.

Mitochondria produce most of the oxygen-free radicals in the body, harmful compounds that "antioxidants" seek to protect against. When mitochondrial function is impaired, the body produces less energy and more "free radicals" are produced. Coenzyme Q10 ("Q10") is a compound central to the process of making energy within mitochondria and quenching free radicals. However, statins lower Q10 levels because they work by blocking the pathway involved in cholesterol production – the same pathway by which Q10 is produced. Statins also reduce the blood cholesterol that transports Q10 and other fat-soluble antioxidants.

"The loss of Q10 leads to loss of cell energy and increased free radicals which, in turn, can further damage mitochondrial DNA," said Golomb, who explained that loss of Q10 may lead to a greater likelihood of symptoms arising from statins in patients with existing mitochondrial damage since these people especially rely on ample Q10 to help bypass this damage. Because statins may cause more mitochondrial problems

over time- and as these energy powerhouses tend to weaken with age – new adverse effects can also develop the longer the patient takes the statin drugs.

"The risk of adverse effects goes up as age goes up*, and this helps explain why," said Golomb. "This also helps explain why statins benefits have not been found to exceed their risks in those over 70 or 75 years old, even those with heart disease." High blood pressure and diabetes are linked to higher rates of mitochondrial problems, so these conditions are also clearly linked to a higher risk of statin complications, according to Golomb and co-author Marcella A. Evans, of UC San Diego and UC Irvine Schools of Medicine.*

Dr. Golomb gos on to describe how the statins may have caused these neurological problems. Statins lead to dose-dependent reductions in coenzyme Q-10, a key mitochondrial respiratory chain defect.

This is just the tip of the iceberg when it comes to the side-effects linked with statin drugs. Golomb also indicated in her studies some of the additional risks linked to statin drug use. Golomb states:

An array of additional risk factors for statin AEs (adverse effects) are those that amplify or reflect mitochondrial or metabolic vulnerability, such as metabolic syndrome factors, thyroid disease, and genetic mutations linked to mitochondrial dysfunction. Converging evidence supports a mitochondrial foundation for muscle AEs associated with statins, and both theoretical and empirical considerations suggest that mitochondrial dysfunction may also underlie many non-muscle statin AEs. Evidence from RCTs (randomized controlled clinical trials) and studies of other designs indicates existence of additional statin-associated AEs such as cognitive loss, neuropathy, pancreatic and hepatic dysfunction, and sexual dysfunction."

Statin drugs also have a blood thinning effect. This is beneficial for reducing the chance of stroke. However, when the statin drug is taken with other blood-thinning drugs, like Coumadin, the effect of the blood-thinner can be increased. This can cause the blood to become too thin. While strokes can be caused from blood clotting or blockages in the artery, they can also be caused by excessive bleeding. Statin drugs and bleeding strokes have been linked together when they are taken in conjunction with another type of blood-thinning drug.

The sad part is that statin drugs aren't as easy to get away from as one might think. Statins are commonly hidden in medications that are known as "combination medications." Common statin medications include Advicor, Caduet, and Vytorin. These combination statin drugs interact with other medications that the patient might be taking.

Often, the interaction between statin drugs and other drugs may have a beneficial result, but they create a higher risk of other problems. For this reason, some combination drugs should not be used long-term.

Before putting a patient on one of these combination drugs, the doctor might say, "Well, we are going to switch you over to Caudet" after a patient expresses concerns about side-effects that they have experienced with statin drugs. If drug-makers could find a way, I think they would find a way to put these drugs in your cereal or drinking water.

While there are benefits to taking statin drugs, often the risks greatly outweigh the costs. This is true, even for combination statin drugs. The reality is that, despite whatever a physician may tell you, there are serious risks to taking any medication, including combination statins. These risks are the same in combination drugs as they are in regular statin drugs. In fact, Golomb indicates that combinations of drugs can sometimes increase the negative, unwanted side effects.

Chapter 3:

The Scary Epidemic – Physician Denial of Statin Complications

The National Health and Nutrition Examination conducted a study based on patient data between 1999 and 2004. According to this study, an estimated 33.5 million older Americans (men older than 50 and women older than 60) currently take a statin drug prescribed by their doctor. That means that if you're over 45, there's a 22 percent chance you're taking a statin medication to lower your cholesterol levels, according to Consumer Reports. http://www.consumerreports.org/health/resources/pdf/best-buy-drugs/StatinsUpdate-FINAL.pdf And that adds up to big dollars. Pfizer, the manufacturer of this drug has earned $12.9 Billion on their super-statin, which outsells any of the competition for more than twice the amount

The cost to Americans without health insurance is $182.46 for 50 pills at Costco, one of the least expensive pharmacies in the USA. Some of these patients are unaware of the serious side effects that often accompany statin use. Ed V. was one of those patients.

A retired linesman for a telephone company, Ed was 59 years old when he began experiencing numbness in his toes. His pain eventually grew to include the balls of his feet and his arches. By the time he came to see me, he had been suffering for about three years and had developed restless leg syndrome; a common symptom of neuropathy.

Ed was a fit and active man. He exercised, played golf and walked. He also held down a part-time job as a bus driver. The only risk factor he had for peripheral neuropathy was years of statin drug therapy, prescribed by his physician.

When he came to our office, we examined him and discovered that he had all the tell-tale signs of small fiber neuropathy. We developed a plan of care that could help him overcome his neuropathy and gain the feeling back in his feet. However, Ed had some decisions to make.

Our care plan builds up the nerves, which helps them work better. If he continued taking the statin drug, his chances of long-term success would be diminished. Understanding this, Ed made the decision to stop taking his daily dose of statin and underwent therapy in our office. Within four visits, his legs were calm at night and the numbness in his feet had disappeared.

Ed was a typical example of how statin drugs can impact the body. Sadly, it never occurred to Ed or his doctor that the drug he was taking to lower his cholesterol was responsible for the discomfort he was experiencing in his legs and feet.

Ed V. is not the only patient that is unaware that statin drug use can create serious health problems, including neuropathy. A study conducted by researcher Beatrice Golomb, M.D, Ph.D., at the University of California in San Diego found that many doctors fail to ask their patients if they are experiencing any of the known side effects. [Ref. 2] What's worse is that many physicians actually ignore the signs of potentially harmful effects of statin drug use when their patients voice concern.

Golomb's study analyzed 650 adult patients that were taking statin drugs. In this study, Golomb documented the perception that these patients had about their doctor's reaction when they expressed concern about the common side effects they were experiencing during their use of statin drugs. These patients reported that their doctors were more likely to deny, rather than affirm, the possibility that there was a connection between the statin and their symptoms.

Before Golomb's studies were concluded, countless patients who experienced statin side effects had their symptoms ignored by their doctor. Doctors either didn't understand the side effects, didn't believe the symptoms could be a result of the drug, or worse, didn't care. Golomb told the Washington Post, "Person after person [told] us that their doctors told them that their symptoms, like muscle pain, couldn't have come from the drug." [Ref. http://www.pharmalot.com/wp-content/uploads/2007/08/drug-safety-2007-physician-response.pdf] In some cases, doctors simply refuse to accept that the statin drugs they are prescribing might be causing some serious problems and side effects for their patients .

In the abstract of this study, Golomb wrote:

"Rejection of a possible connection was reported to occur even for symptoms with strong literature support for a drug connection, and even in patients for whom the symptom

met presumptive literature-based criteria for probable or definite drug-adverse effect causality. Assuming that physicians would not likely report ADRs in these instances, these patient-submitted ADR reports suggest that targeting patients may boost the yield of ADR reporting systems," [Ref. http://www.ncbi.nlm.nih.gov/pubmed/17696579]

It was concluded in this study that, since physicians often ignored these symptoms and their severity, low reporting rates likely contribute to delays in the identification of adverse drug reactions.

People trust their doctors, but for 87 percent of the patients in this study, their trust in their physician was betrayed. An astounding 87 percent of patients in the study said they told their doctors about their symptoms, only to be ignored or told that their symptoms were not related to the drugs when they expressed concern about the possible connection between their symptoms and statin drug use.

Dr. Golomb's research received quite a bit of media attention and people quickly started talking about the fact that, when patients mentioned their statin drug side effects to their doctors, they were sometimes ignored or their symptoms were disregarded. This led to many different articles published on the subject, as well as additional studies.

An article, published in the *Diabetes Journal* in December of 2007 highlighted what medical writers like Dr. Golomb were saying about this newly discovered aspect of the drug. This article also discussed four of the most common myths about statin drug use. [Ref. 4 http://docnews.diabetesjournals.org/content/4/12/1.1.full]

1. *Myth: Muscle-related side effects are limited to muscle pain. -- According to the recent National Lipid Association Statin Safety Task Force report, statin-related muscle symptoms (generally referred to as myopathy) can include muscle pain as well as more general soreness, weakness, and cramps. 'The statin association is often made with bilateral and proximal muscle pain, but there can definitely be exceptions to listen to.' LaRosa says.*
2. *Myth: Only an elevated creatine phosphokinase (CPK) test indicates a statin-related muscle effect. — 'Some physicians may assume that if a patient's CPK is not elevated, then there cannot be a statin*

association' Golomb says. 'But this is repeatedly refuted in the literature. Patients with minimal or no CPK elevation can still have reversible myopathies that are impacting their quality of life.'

3. *Myth: Statins only help – they do not hurt – cognitive ability. – While several high-profile studies have linked statin use to lower incidence of dementia, to date they have been uncontrolled observational studies. 'That sort of study design has repeatedly led people astray,' Golomb says. 'It's the same design that showed estrogen use dramatically lowers rates of Alzheimer's disease. Of course, in the high-quality matched studies, hormone replacement therapy actually increased risk of dementia.' Two randomized, double-blind, placebo-controlled trials found that statins may cause cognitive decrements – such as loss of memory and concentration – in some patients taking statins. However, two other randomized trials found no such effect. 'What makes us believe these negative effects can be statin-related is that, for many people, they reverse then recur with statin re-challenge,' Golomb says.*

4. *Myth: Side effects only surface during the early stages of treatment. – 'We've seen that side effects can develop long after patient has started on statins,' Golomb says. Increasing age raises the risk for adverse effects, as do several medications commonly prescribed for aging adults. These medications can adversely interact with statins.*

Patients rarely hear of the side effects and risks when they are given a prescription for statins. Sometimes, when they begin to experience common statin-related problems, they are often left to fend for themselves. Doctors are often so eager to prescribe them that they only give patients the positive. This leads to patients who take them, expecting no downsides. However, there are patients who recognize the side effects of statin drug use. When they don't get the right response from their doctor, they persevere and take their care into their own hands. Adele M. was one of those patients.

At age 72, Adele began experiencing painful, burning feet and a cold sensation in her toes that would not go away. These strange sensations seemed worse in the evening and she could not relax. She developed restless leg syndrome.

When she tried to sleep, the constant urge to move her legs kept her awake. Adele had something in her favor, however. She was a retired prescription drug representative with some medical training and she liked to do research.

When she started experiencing the symptoms, she went online, did some research, and came to the conclusion that her symptoms were caused by peripheral neuropathy, a side-effect of the statin drug her doctor had prescribed to her. Adele went to her doctor and said, "I think I have peripheral neuropathy." According to Adele, her self-diagnosis only aggravated her doctor.

He sent her to a neurologist who prescribed B-12 vitamins and Gabapentin to her, both of which are commonly prescribed to treat peripheral neuropathy. These drugs also came with side effects that she didn't like. She said that she felt "loopy" and "not herself." To top it off, the new medications didn't seem to relieve her of the original symptoms she had complained about.

By the time Adele came in to our office, she had been on statins for so long that she could not remember when she had started them but she thought it had been at least ten years. She was at the end of her rope. She wanted relief and needed to get off of those drugs. Fortunately for Adele, within five visits to our office, her feet returned to their normal pink color and, for the first time in years, she was completely pain-free and finally statin free.

George S. was another one of my patients. When George first came to see me, he was 88 years old. He was pretty certain that he had neuropathy. His feet burned all of the time and he had trouble keeping his balance. He had recently taken two falls and his wife was starting to worry about him.

Every step was difficult and painful, forcing him to use a walker. The pressure of shoes had become so unbearable that he could only wear slippers. His legs were extremely swollen and a dark discoloration extended down his legs and all the way down to his toes. We did an exam on George and confirmed that he did, indeed, have a serious case of peripheral neuropathy. We began a treatment plan and he began to improve.

Within ten visits, he was able to abandon the walker for a cane. He followed our program diligently and his old, cheerful self eventually returned. George finally got the mobility and quality of life back that prescription statin drugs and robbed from him and his wife. She, more than anyone, appreciated the improvement. She no longer worried about George, or his health.

I see countless cases like George, like Adele, and like James. Many of them, the neuropathy sufferers, take statin drugs. Many have tried to talk to their doctors about their neuropathy. Many have asked if their neuropathy could be related to their statin drug use. Yet, their physicians continue to ignore their cries for help.

My experiences with my patients match that of Golomb's studies. Golomb expresses that there is a serious need for physician sensitivity when it comes to the potential negative side effects of statin drugs. They need to learn to listen to their patients. They need to hear their patients and compare their patients' complaints with that of known adverse effects of statin drug use. They also need to take precautionary measures when serious side effects are expected.

Until this happens, I have a responsibility to my patients. I have a responsibility in the medical community. I cannot sit idly by while millions of people suffer from statin induced neuropathy. Statin users deserve to know the truth.

Chapter 4

Statin Drug Use and Cancer

As if neuropathy, muscle pain, dizziness, and memory loss weren't enough risks behind statin drug use, studies indicate that there may be a link behind cancer and statin drug use. While many research papers refute this possibility, there are other studies that indicate that these studies are inaccurate.

In 2006, the American Medical Association Journal [Reference http://jama.ama-assn.org/content/295/1/74.full.pdf] published a paper about statin drug use and cancer. It was based upon a meta-analysis conducted by the

University Of Connecticut School Of Pharmacy in Storrs, Connecticut.

The study wanted to investigate whether or not statin drugs could prevent cancer. The study concluded that "statins have a neutral effect on cancer and cancer death risk in randomized controlled trials." Additionally, the study concluded that "no type of cancer was affected by statin use and no subtype of statin affected the risk of cancer." The article discussed the lack of cancer-reducing capabilities of statin drugs outlined in the study.

> *Statins do not reduce the risk of cancer, finds a review of several long-term studies published today.*
>
> *The findings contradict previous studies, which suggested that the cholesterol-lowering drugs reduce cancer risk. They are based on 26 randomized controlled trials of statins and cancer incidence, or cancer death, including a total of 86,936 participants.*
>
> *Dr C. Micheal White and colleagues at the University of Connecticut and Hartford Hospital, Hartford, USA, report the findings in today's issue of the Journal of the American Medical Association.*

They write: 'In the trials, statins reduced the risk of a first myocardial infarction [heart attack] and overall mortality. With long-term follow-up and collection of cancer data in a majority of studies, insight into the risk of cancer among statin-niave persons and statin users can be derived.'

'In our current meta-analysis, statins did not reduce the incidence of cancer death.'

*JAMA 2006;
295: 74-80*

Another article, published by the Journal of American College of Cardiologists on Aug. 22, 2008 examined the relationship between lower cholesterol levels and higher risks of cancer. In a study that was designed to assess the risk of cancer in statin drug patients, researchers found that individuals taking placebo or statin drugs had a higher risk of cancer when their cholesterol levels were lower. However, researchers determined that, for a given cholesterol level, statin takers, overall, were not found to be at an increased risk of cancer compared to those taking placebo.

The study led to many press stories that claimed that statin drugs do not cause cancer. The article went on to question whether or not low cholesterol levels could be the actual culprit. [Ref. http://www.medscape.org/viewarticle/579456, http://content.onlinejacc.org/cgi/reprint/52/14/1141.pdf]

> *"And let's imagine that through their cholesterol reducing capacity statins can, therefore, enhance cancer risk. Well, it likely takes some time for statins to reduce cholesterol and almost certain even longer for that to manifest as full-blown, diagnosable cancer. In other words, the statin trials may simply have not gone on long enough for the long-term effects on cancer to be properly assessed. Looking at individual studies within the review, we find that some studies show no statistically significant increased risk of cancer in statin takers. Some studies even showed a reduced risk of cancer. But, and this is important I think, some individual studies did show an increased risk of cancer in those taking statins. The results of these studies can, of course, be diluted by the other studies, but they are still there, and some*

would argue that their presence casts a pretty ominous shadow over statin and perhaps cholesterol-reducing therapy generally."

The Times, an English newspaper also published an article in 2007 that discusses the relationship between lower cholesterol and cancer. The article does not point the finger at statin drugs so much as the effect that statin drugs have on the body; their ability to lower cholesterol. [Ref. http://www.timesonline.co.uk/tol/life_and_style/health/article2127605.ece]

> *It is not clear whether the cancer cases are caused by the drugs, or are a consequence of the low levels of 'bad' LDL cholesterol produced by taking them.*
>
> *The result, which amounts to one extra case of cancer for every 1,000 patients treated, surprised the researchers who discovered it. They were looking for new evidence on the known side-effects of statins on the liver and muscle wasting.*
>
> *'This analysis doesn't implicate the statin in increasing the risk of cancer,' said the study leader, Professor Richard Karas, of Tufts*

University School of Medicine, in Boston. 'The demonstrated benefits of statins in lowering the risk of heart disease remain clear. However, certain aspects of lowering LDL with statins remain controversial and merit further research.'

What's more, the study written about in this article discusses the fact that aggressive cholesterol-lowering treatment may have even more of a negative effect on the body. It was found that, not only was there a possibility of cancer but that serious damage to the liver can be caused as well.

One would wonder exactly how out of touch the FDA can be. While there are no definitive results about the relationship between statins and their ability to cause cancer, there are numerous sources that express the need for more research. And really, if cancer is a possibility, why are these drugs even being prescribed?

Maybe we have all become so dependent on a magic pill that we are willing to take whatever is given to us without even bothering to question. Maybe the pharmaceutical companies have picked up on this fact. No one really knows why all of these side effects are being ignored. The truth is that it doesn't really matter. What matters is that my patients learn that there is a better way to manage their cholesterol so that they can get back their life, reduce their chances of experiencing side-effects from statin drugs, and regain their mobility. This is essential to the quality of life for many of my patients as well as other neuropathy sufferers around the world.

Chapter 5

The Cholesterol-Statin Story

How did we get to this point? Which came first; the drug or the disease? Why is it that approximately 24 percent of the American population is taking a statin drug or statin drug combination?

Many statistics have shown that, before the 1920's, cholesterol-related conditions were a rare occurrence. Diet primarily consisted of grains, fruits, and vegetables. Meat and other animal products were rarely consumed. This is because the average American family was too poor to afford these types of food.

High blood pressure, heart attack and stroke mostly existed among the wealthy. These families were able to afford "finer foods" like bacon, chicken, and roast. It was not until animal products like milk, eggs, cheese and meat became more affordable that heart disease and other related conditions started to take over America.

Thirty years ago, hypercholesterolemia, the medical term for the presence of high levels of cholesterol in the blood, was a newly discovered disease that was starting to become more common. Few individuals were actually treated for it. Typically, to receive any type of high cholesterol diagnosis and treatment, a patient had to be a middle-aged man with cholesterol levels that were above 240. He would also have to have other risk factors like smoking or obesity.

Cholesterol became a mainstream topic of conversation on March 26, 1984, when Time Magazine ran a cover story entitled, "Cholesterol." The cover featured two fried eggs on a plate resembling eyes and a downward curving slice of bacon, resembling a frown. "And now the bad news," the cover warns. "Cholesterol is proved to be deadly and our diets may never be the same..." This is how the article begins. It goes on to discuss results of a government study that examined the link between a fatty diet, cholesterol levels and heart disease.

For some readers, the article was an introduction to cholesterol. It was the first that they would hear of cholesterol-lowering drugs. The writers of the article indicated that medication may be necessary to lower cholesterol; that a low fat diet did not lower cholesterol.

In the exact same year that the New York Times published their article on cholesterol, the Cholesterol Consensus Conference changed the parameters of what was considered high cholesterol. These new parameters included anyone with levels over 200. Suddenly, millions of patients were faced with the battle of reducing their cholesterol. At first, this was through diet and exercise. However, cholesterol-lowering drugs were already in the research phase.

In 1987, just a few years later, statins were given the pre-market approval. This allowed for scientific and regulatory review of statin drugs. Within just a few more years, the cholesterol-lowering drugs had become well-established in pharmacies nationwide. In 2004, the acceptable cholesterol level was lowered again. It was now down to 180.

These new guidelines were similar to the idea of me telling a patient, "Okay, everybody wants to come to the chiropractor, that's fine but, you can't leave my office until you can bench press 300 pounds and run a two-minute mile." That would create a situation where patients would never reach the point where they can stop coming to the chiropractor. They will never be perceived healthy enough so the goal becomes impossible to reach.

Almost anyone can receive a statin prescription to lower their cholesterol levels, whether it is needed or not. Any man or woman with cholesterol levels above 180 is routinely written a prescription for statin drugs. If a patient has had a heart attack, they are prescribed a statin drug as well, even if their cholesterol levels are below 180. This is based on an assumption that the stroke is related to high cholesterol. Testing may not even be performed to confirm this theory.

Once the guidelines were changed, high-risk people – those who have had a heart attack, have diabetes, chest pains, or have had surgery to clear blocked vessels – were told to achieve an LDL level of 70, instead of the usual 100. This new threshold would push down, even further, the patients who were considered to have cholesterol issues, even if they weren't at risk for heart disease.

Diabetics are also commonly treated for high cholesterol, even if their cholesterol levels are low. Sadly, the cholesterol treatment ignored the studies conducted, which are currently under review. Three of the four studies conducted on patients with diabetes showed that there was no significant benefit with the increased use of statins. Taking the finding as at face value, one death was prevented each year out of the 250 diabetic patients that were treated with a statin drug. However, these patients are at a higher risk from peripheral neuropathy, simply because of their condition. Add statin drug use to the mix, and their chances of developing neuropathy are dramatically increased. This is extremely worrisome!

Among the other groups of patients receiving cholesterol-lowering therapy with statin drugs are people who are overweight, smoke or have a narrowing of the arteries. People with a family history of heart disease or cholesterol may not display any problems with cholesterol. However, based upon this family history, they too can be prescribed statin drugs. Individuals who are inactive or in poor general health may also receive cholesterol-lowering therapy.

In some cases, any man older than the age of 55 or any woman older then the age of 65 are prescribed statins for no other reason than their age. This is done, despite the fact that studies have proven that statin drug side-effects are more common and more severe in elderly patients. This is done, despite the fact that the studies indicate that older patients receive less benefits from statin drug use than younger patients.

Many new patients receiving cholesterol-lowering treatment are unlikely to see any benefits from the therapy. In fact, there are no studies that indicate that women with or without heart disease benefit from statins drug use. One study looked at elderly patients over the age of 70. This study did not show a significant reduction of heart deaths, but it did show a significant increase in cancer deaths. The odds are that the report dismissed the findings of the study by combining the cancer death data with other studies of younger patients who were at a lower risk for developing cancer from the drug.

So why is all of this happening? Well, while much can only be speculated, there is one truth about the new guidelines. These new guidelines added millions more to the ranks of statin patients, establishing a new steady market for drug makers. Is this simply a coincidence? Who really knows, but the drug manufacturers are now making a fortune off of these new lower cholesterol parameters.

Having millions of people convinced that they needed a statin prescription to lower cholesterol and to lower their chances of heart disease meant good news for the companies who manufactured the drugs and a host of new drugs were developed to lower cholesterol levels. Among the new drugs were Lipitor, Lescol, Mevacor, Altoprev, Pravachol, Crestor and Zocor.

The medications targeted the liver, thwarting the formation of cholesterol. Most were effective at lowering LDL, but some also lowered triglycerides and raised HDL. However, what you won't hear is that out of 100 patients, only one will receive real, life-saving benefits from the use of statin drugs. This is found in the fine print, the part that almost no one reads. There is a reason that this information isn't easily accessible or commonly disclosed to those with high cholesterol.

With more than 18 million people taking the drugs, statin drugs raked in $21.6 billion in sales nationwide during 2006. Now more than 25 million Americans take statins regularly, accounting for the majority of statin drug sales. This figure was shown in a great article called "Statins with your burger?' Doctors want heart pills on menu" where a UK reporter discusses McDonald's, Burger King and other fast food outlets should offer diners free drugs to compensate for the risk of heart disease, cardiologists proposed. [Reference http://www.sodahead.com/united-states/statins-with-your-burger-doctors-want-heart-pills-on-menu/question-1460163/]

Cholesterol fighter Lipitor held the title "best-selling drug" for a few years, and has been a major source of income for the world's biggest drug company, Pfizer. Lipitor (atorvastatin) was released in 1998, and by2006i t had reached peak sales of $12.9 billion, accounting for 27% of the company's revenue. In 2010, with $10.8 billion in sales, Lipitor still accounted for 15.8% of total revenue, even with the addition of Wyeth's operations. [ref http://www.dailyfinance.com/2011/02/27/top-selling-drugs-are-about-to-lose-patent-protection-ready/]

These new recommendations helped increase the sales of statin drugs by 20 percent. The drug-makers realized, "If I tell everybody that they need to take this drug, I am going to make more money." Unfortunately, the drugs often prescribed to help regulate cholesterol actually end up creating problems. Yet, all of this was ignored, more people were now receiving statin therapy, and the prescription drug companies were now making a killing off of it.

In May of 2002, Dr. Julian Whitaker filed petitions with the U.S. Food and Drug Administration (FDA), calling for black box warnings on all statin drugs. These warnings would be similar to the warnings used in Canada for statins.[ref http://www.lef.org/magazine/mag2002/aug2002_legalnotes_01.html] His petitions stated that coenzyme Q10 depletion, as a result of statin drug use, can put patients at risk for impaired myocardial function, liver dysfunction, and myopathies, including cardiomyopathy and congestive heart failure – the very conditions that it claims to protect from. The FDA has yet to respond.

Dr. Julian Whitaker has also published literature on the use of statin drugs for lowering cholesterol. Whitaker states that "statins do not save lives." [ref http://www.whitakerwellness.com/content/article/What-You-Don-t-Know-About-Statins-Can-Hurt-You.html] In fact, Whitaker even discusses a

New England study that claims that there was a 22 percent reduction in heart disease deaths in a group of patients that were on a cholesterol-lowering treatment plan using Lipitor. However, this study did not disclose the fact that there were a number of deaths from statin-related side effects. This fact basically ruled out any lives that were saved by the cholesterol-lowering drug.

While the longer term implications of coenzyme Q10 deficiency are serious, using LDL cholesterol as the primary (and often the only) predictor of heart disease, as set out in the National Institutes of Health's Expert Panel on Detection, Evaluation, and Treatment of High Blood Cholesterol in Adults, may be entirely inappropriate for a number of reasons, according to cardiologist Stephen T. Sinatra, MD, FACC. [Sinatra ST, MD. "Is cholesterol lowering with statins the gold standard for treating patients with cardiovascular risk and disease?" (editorial), Southern Medical Journal, 3/1/2003.] The report outlines the following:

- A threefold increase in the number of Americans given prescriptions for statin drugs – a move that would increase the sales of statin drugs by about 20 percent annually.

- The oxiditaion of LDL cholesterol remains a hypothesis for heart disease, yet cholesterol levels continue to drive treatment.
- Research on omega-3 fatty acids shows improved survival rates for those with cardiovascular disease, independent of cholesterol levels, lending support for reducing inflammation as a way to reduce cardiovascular mortality.
- Statins may play a role in intervention and treatment because of their anti-inflammatory properties – not their cholesterol-lowering properties. In fact, one study indicated that about half of myocardial infarction patients have a normal LDL cholesterol level and that C-reactive protein may be a better predictor for atherosclerosis.
- Compared to patients taking statin drugs, untreated subjects with normal C-reactive protein levels did not have an increased risk of recurrent cardiac events, while those whose C-reactive protein levels were elevated did have a significant risk of fatal coronary events, regardless of their LDL cholesterol levels.

According to Dr. Sinatra, the report rejects other potentially, more accurate predictors of cardiovascular disease, including C-reactive protein and homocysteine levels – both measures of internal inflammation [ref. Ridker PM, Stampfer JM, & Rifai N. "Novel risk factors for systemic atherosclerosis: A comparison of C-reactive protein, fibrinogen, homocysteine, lipoprotein(a), and standard cholesterol screening as predictors of peripheral arterial disease." JAMA, 285:2481-2485, 2001.]. Dr. Sinatra's statements are starting to be heard throughout the medical community. More and more studies are producing the same results – cholesterol levels are not the primary indicator to heart disease. In many cases, it may not be an indicator at all.

Probably the worst part of all of this is that, by talking to patients about diet and exercise, doctors could save four times as many lives than prescribing statin drugs. While the new guidelines do mention lifestyle changes to lower cholesterol and it is recommended that drugs should not be prescribed without advice to exercise, eat more fruits and vegetables, fibers and less fat, so few patients are getting the proper education about natural cholesterol lowering methods. In fact, the Center for the Science in the Public Interest said that the guidelines give only a cursory mention to using natural, rather than drug-induced methods to lowering cholesterol.

Eight out of the nine others of the JAMA recommendations had financial ties to statin manufacturers, including Pfizer and Merck, but that fact was not disclosed when the recommendations were first published. It was the equivalent of getting a bunch of buddies together, handing out some money and saying "let's look at these studies." This group of scientists says, "Oh, we should push this part. Get it published, then everybody has to do it." That's bad news. However, their new guidelines, their push on the administration of statin drugs did not go unnoticed.

Changing Parameters Questioned

In 2004, the new guidelines that were issued by a nine-member panel, including members of the American Heart Association, American College of Cardiology and federal officials added 7 million new patients to the 36 million people already taking statin medications. These new guidelines caught the attention of the Center for Science in the Public Interest (CSPI) which claimed that the new rules were flawed because eight of the panel members were financially tied to the drug makers.

The CSPI accused the National Institutes of Health and the National Cholesterol Education Program panel [ref. http://circ.ahajournals.org/cgi/content/full/106/25/3143] of having been influenced by their connections to drug companies when issuing guidelines for statin use, recommending the drugs for people without high cholesterol, to help them lower their levels. [ref. http://www.cspinet.org/integrity/press/200409231.html] In a press release, the group wrote, "Eight of the nine authors of the July recommendations have financial ties to statin manufacturers, including Pfizer, Merck, Bristol-Myers Squibb, and AstraZeneca – a fact that was not disclosed when the recommendations were first published in the Journal Circulation. Merril Goozner, director of the Integrity in Science project at CSPI, said that those conflicts of interest alone are enough to warrant skepticism of the NCEP findings."

The Center for the Science in the Public Interest is a strong advocate group for nutrition, health and food safety. They have 900,000 members in the United States and Canada and are not sponsored by anybody. The group strives to educate the public about government policies that are consistent with scientific evidence of health and counter industries that have a powerful influence on public opinion and public policies. They study animals, nutrition, everything. This group is the one that successfully lobbied to have food manufacturers list the nutritional information of their food on food labels. They are also the group that fought the food industry in reducing trans-fat in their products.

This group sent a petition to the NIH, asking them to take a closer, impartial look at statins and the relationship between the drug companies, the researchers and the newly lowered cholesterol guidelines. Their petition stated:

> *"When researchers have financial relationships with drug companies, [it] raises flags. When those relationships are concealed, alarm bells start going off....Drug companies want to sell more drugs and they fund studies to further that goal. That's why it is critical that scientists who review industry-funded studies and write clinical practice guidelines on behalf of the government be completely free of conflicts of interest."*

The group also accused the NIH of ignoring the benefits of lifestyle changes on health.

> *"The sad fact is that these lifestyle recommendations are being largely ignored, partly because the 'experts,' many of whom have conflicts of interest through their relationships with statin manufacturers focus ever more attention on lowering cholesterol with expensive drugs....The vast majority of heart disease can be prevented by adopting healthy habits. The American people are poorly served when government-sanctioned clinical recommendations, uncritically amplified by the media, misdirect therapies that may not be scientifically justified."[Ref. http://www.cspinet.org/integrity/press/200409231.html]*

The drug makers also came under fire when critics said that advertisements for Lipitor, featuring Dr. Robert Jarvik, inventor of the artificial heart, misrepresented the medical scientists' credentials, deliberately giving the wrong impression that the medical scientist was touting the drug as a physician.

A congressional committee later asked an investigation of the ads, saying that the advertisements gave the wrong impression that Jarvik, a researcher and inventor, saw patients and prescribed medications, which he did not. In media interviews Jarvik defended the advertisements, stating that he tried to lower his cholesterol through diet and exercise alone for years before his physician prescribed Lipitor for him, which finally lowered his numbers. Although Jarvik touted the advertisements as an "educational campaign," Pfizer later pulled the ads, saying that misrepresentations of the spots were "distractions."

Despite all of the controversy around statin drugs, many doctors are still prescribing them to patients. What's worse is that they are prescribing them to the wrong patients. They continue to ignore the facts. They refuse to educate their patients about the more effective methods of handling the risk of heart disease. The longer this goes on, the richer the statin drug companies get. Even worse, the longer this goes on, the greater the number of patients that are at risk for statin-induced side-effects.

Chapter 6

Should Your Doctor Write a Shopping List Instead of a Prescription?

Lowering Cholesterol Without Statins

The most well-known advertisement for Lipitor touts the medication's ability to lower cholesterol "when diet and exercise are not enough." This leads way to the question: why isn't diet and exercise not enough?

Since high cholesterol exhibits no symptoms, most people don't even know they have the condition until a doctor tests them for it and then advises them that they need medication. Since that Times Magazine article in 1984, there have been a slew of books, documentaries, news articles and stories about the dangers of a high fat diet. Certainly most consumers have realized that a healthy diet leads to good health. So what is the root of the problem? Why do people continue to consume unhealthy foods that they know could be detrimental to their health?

For most people, the cholesterol problem begins in the grocery store, shopping for food. In trying to find the fresh ingredients to make homemade foods, shoppers encounter hurdles and road blocks. There is an overabundance of pre-packaged foods and frozen meals. Fresh foods are sometimes more expensive so putting together a home-cooked meal can seem more costly. What's more, the average American feels entirely too busy to spend the time preparing a meal from scratch. Buying a frozen or pre-packaged meal can seem like the cheaper alternative as well as the most time efficient.

What most shoppers don't realize is that the average pre-packaged meal comes loaded with a tremendous amount of chemicals, fat, and salt; much more than their made-from-scratch counterparts. Therein rests the start of a problem.

Another thing that most shoppers don't realize is that most grocery stores are set up to create these hurdles. In the center of the store, the very first area you often enter, are all of these convenient food items. To purchase the foods that are best, you have to extend to the outer rim of the grocery store. Healthy food items are harder to locate and get to.

Additional problems come in when you consider the fact that the American diet is one of the few that bases its meals on meat products. Nearly every American dish includes a meat component. Meat, of course, is high in both cholesterol and fat. Add in the pre-packaged convenience foods that so many Americans opt for because they are simply "easier" and we are starting to see where all of the confusion comes in.

So when asking our question again: Why isn't diet and exercise enough – is it truly because it isn't enough or is it because patients aren't being properly educated about their diet? Can the right diet, the right education, and the right information about diet make a difference? In most cases, the answer is yes.

So what is a healthy diet? When asking this question, you are likely to get a plethora of different opinions. There is, of course, the

American Standard Diet. There are raw diets. There are diets for almost everything. However, there are only a few diets that have been tested on individuals with high cholesterol and a risk for heart disease.

Psyllium Husks

Psyllium husks are nothing new. They have been used as a dietary supplement in Europe and parts of Asia for decades. In 1998, the FDA approved the use of psyllium husk as a dietary supplement in the United States.

Psyllium is derived from the more than 200 species of Plantago plants and the fiber is obtained primarily from the seeds. In the United States, psyllium is usually found in high-fiber cereals and some baked goods; particularly those that are touted as gluten-free.

Consuming the fiber in psyllium husks is the most effective way to include it in the diet. This increases the amount that is absorbed. However, it is also sold in a supplement form.

Soluble fiber forms a gel when mixed with liquid. The FDA recommends that people consume between 22 and 30 grams of fiber a day. Psyllium seed husks provide between three to 12 grams of soluble fiber and, when included as a fiber in a low-fat, low cholesterol diet, it could reduce the risk of heart disease according to a report by the FDA.

Many other studies have shown that the high fiber in psyllium husks is effective in lowering cholesterol, naturally. More than 18 PubMed studies have highlighted its effectiveness and few have disputed the plant's benefits. Yet, these plants are not widely used nor are they widely discussed by physicians when talking to their patients about lowering their cholesterol.

Thanks to a successful advertising campaign, just about everyone has heard of the statin drug Zocor. It is safe to say that this drug is well-known. Yet, not everyone has heard of psyllium husks, despite the fact that they have been proven to be nearly as effective at lowering cholesterol as pricey-statin drugs, and they are much cheaper. They also have fewer side effects.

A study was conducted at the School of Pharmacy, University of California, San Francisco, examining the impact that psyllium husks have on hypercholesterolemia. The study stated that:

> *In summary, within a controlled study situation, psyllium seems to be effective in lowering total and LDL cholesterol by 4-8% and 6-13%, respectively.....NCEP recently*

has stressed diet therapy as a first-line primary intervention in patients not a high risk from multiple risk factors or very high LDL cholesterol concentrations. [Reference http://www.sciencedirect.com/science/article/pii/000291499290980D]

Another article published in the Indian Heart Journal by James Adams discusses the effect of psyllium husks on hyperlipidemic patients:

Ispaghula husk (psyllium) is nearly as effective as simvastatin for improving the lipid profile of hyperlipidemic patients.

It is also much cheaper and has fewer side effects than does simvastatin, report investigators from the Government Medical College in Amritstar, India.

The investigators randomized 60 hyperlipidemic patients into two groups of 30. One group received 3.5 grams of ispaghula husk twice a day and the second group received 20 milligrams of simvastatin each day. Treatment continued for 12 weeks, and lipid profiles were evaluated at 0, 4, 8, and 12 weeks.

Results showed that total cholesterol decreased by 15.8 percent and low-density lipoprotein (LDL) cholesterol decreased by 22.97 percent among patients taking ispaghula husk.

Triglycerides decreased by 20.89 percent and high-density lipoprotein (HDL) cholesterol increased by 10.69 percent in these patients.

Among patients taking simvastatin, total cholesterol decreased by 24.15 percent, LDL cholesterol decreased by 36.08 percent, triglycerides decreased by 20.47 percent and HDL cholesterol increased by 11.4 percent.

Ispaghula husk is an effective and well-tolerated dietary adjunct for patients with hyperlipidemia and should be considered before initiating conventional drug therapy, the investigators conclude. [Ref. http://www.pslgroup.com/news/content.nsf/news/8525697700573E1885256C860060D1E2]

In addition to increasing fiber in the diet, psyllium husks can also aide in digestion. They can also act as a laxative or lower blood glucose levels. Some studies even suggest that psyllium husks may reduce the risk of acute pancreatitis.

	• Average statin drug	• Psyllium husk
Lowers Cholesterol	• Yes	• Yes
Improves Digestion	• No	• Yes
Budget-Friendly	• No	• Yes
Cognitive Side Effects	• Yes	• No
Muscle Weakness Side Effects	• Yes	• No
Neuropathy Side Effects	• Yes	• No
Links to cancer	• Yes	• No

However, the studies on pancreatitis still needs more evidence and the safety of these therapies needs to be established.

All of the studies have shown that psyllium husks are nearly as effective at treating cholesterol problems as statin drugs. They are less expensive.

What's more, they don't create the same damaging effects on the nervous system or the muscles. They don't cause cancer or liver dysfunction.

When you compare statins and psyllium husks, their benefits, cholesterol-lowering capability, and side effects side-by-side, it looks something like this:

Psyllium husk vs. Statin drugs

	• Total Cholesterol Decrease	• LDL Decrease
Psyllium Husk Group	• 16	• 23
Statin Drug Group	• 24	• 33

Psyllium supplementation, however, is just part of the overall plan to lowering your cholesterol naturally. Following a complete cholesterol-lowering diet is the best way to lower your cholesterol and reduce the chances of heart disease, without the use of medication.

One such diet has received very little attention, yet it has been proven to be very effective. This diet, referred to as the "portfolio diet," cannot be found in any bookstore but we will review it here. One has to wonder why this diet is not being prescribed to every patient, before they are placed on statin drugs.

The Lean Plate

A cholesterol busting diet can be just as effective as anti-cholesterol pills and there are no side effects and very little cost. David Jenkins, the director of clinical nutrition at St. Michael's Hospital in Toronto, is a vegetarian. In 2006, his research showed that a vegetarian diet works as well as cholesterol lowering drugs. His research was later published in the Journal of American Medical Association. [Ref. http://www.ncbi.nlm.nih.gov/pubmed/12876093] He is the same nutritionist that created the glycemic food index that changed the face of diabetes and food.

Jenkins conducted a test with three different groups.

- Group 1, the control group, consumed a low-fat diet.

- Group 2 was given the same diet in combination with a generic statin drug, Lovastatin.
- Group 3 was given a combination of cholesterol fighting foods including plant sterols, cholesterol-lowering margarines, soy proteins, sticky fiber (psyllium), fruits, vegetables, oats, legumes, and almonds.

The control group, by simply changing their diet, lowered their cholesterol by 8 percent. The statin group lowered their cholesterol by 31 percent. The third group, the dietary group, lowered their cholesterol by 29 percent. While this was 2 percent higher than the statin group, the dietary group also experienced a drop in their C-reactive proteins. As discussed earlier, this may have more to do with heart disease than cholesterol itself. Additionally, C-reactive proteins are a definitive risk for all pain and achiness. The dietary group was the only group to see a change in their C-reactive proteins. [ref. Jenkins D.J., Kendall, C.W., Marchie, A., et. al. 2003. Effects of a dietary portfolio of cholesterol-lowering foods vs Lovastatin on serum lipids and C-reactive protein. JAMA. 290(4): 502-10]

Dr. James Anderson, a professor of Clinical Nutrition and Obesity at the University of Kentucky also advocates a heart-healthy diet. This diet restricts fat and adds more sticky fiber as well as some soy protein. According to Anderson, a basic heart healthy diet has three key ingredients:

- Soy – includes soy-based meat substitutes like soy burgers, soy hotdogs, soy cold cuts, and soy milk
- Psyllium – like Metamucil, plant sterols, Benecol, Take Control, and Smart Balance.
- Nuts – includes almonds, walnuts, pistachios, and others.

Anderson's typical diet is described on the website *Vegetarians in Paradise.* (Ref http://www.vegparadise.com/)

> *A typical portfolio breakfast includes hot oat bran cereal, soy beverage, strawberries, psyllium, oat bran bread with enriched plant sterol margarine, and double-fruit jam.*
>
> *Lunch featured spicy black bean soup and a sandwich that included soy deli slices, oat bran bread, enriched margarine, lettuce, tomato, and cucumber.*
>
> *Dinner highlights were a tofu bake and ratatouille with ingredients like eggplant, onions, and sweet peppers. Pearl barley and vegetables like broccoli and cauliflower completed the meal. Three snacks daily included soy beverages, almonds, psyllium, and fresh fruits.*

The Portfolio Eatin Plan recommends:

- *30 grams (about one ounce) of almonds. (Equivalent to about 23 almonds or 2 handfuls.)*

- *20 grams (less than one ounce) of viscous fiber from foods such as oats, barley, and certain fruits and vegetables.*
- *80 grams of soy protein from foods such as tofu, soy meat alternatives, and soy milk.*
- *2 grams (.064) of plant sterols from foods such as Benecol or Take Control*

Those following the dining plan must also consume daily fruit and vegetable recommendation of five to nine servings and refrain from dairy, meat, poultry, fish and eggs. If they eat an egg product, it should be limited to egg substitutes or egg whites. At least two meals a day are to be vegetarian and those who consume meat should only eat lean cuts.

While this diet may seem difficult to the average consumer, it is actually very simple. And, in many other parts of the world, Venice for example, fresh fruit stands are everywhere. Incorporating these foods into their daily diet is routine.

The basic rules are to avoid trans-fats, avoid refined sugars as much as possible, take cod liver oil, and to include the above foods as well as a large variety of fruits and vegetables into your daily diet.

According to the book, "Lowering Cholesterol by Eating a Soy Rich Diet," the eating plan (Ref http://www.soyfoodsillinois.uiuc.edu/PortfolioPlanSoy.pdf), the eating plan would look something like this:

- *Breakfast: a grapefruit, followed by porridge made with soy milk.*
- *Snack: a small bag of dried fruit.*
- *Lunch: baked potato with a bit of Benecol spread, baked beans and salad with a pear for desert.*
- *Mid-afternoon snack: a handful of almonds and a yogurt drink.*
- *Dinner: chicken and tofu stir-fry and a baked apple with soy milk custard.*

Portfolio Diet Recipes

Sometimes, consumers find the diet to be difficult simply because they feel like their diet is boring. However, there are many delicious vegetarian and vegan dishes that can be enjoyed. Many of them are very simple to prepare and only take a few minutes.

Recipes for vegetarians are often very similar to recipes for meat eaters. Foods like lasagna can be prepared using tofu and eggplant. Enchiladas can be filled with black beans, corn, salsa, and tofu. Your favorite vegetables can be chopped up and placed in a whole wheat tortilla or whole wheat pita bread for lunches. Meat-free salads with oil and vinegar dressing can also make great lunches or summertime meals.

The following are a few recipes that can be used in the portfolio diet.

Tortilla Soup

1 can kidney beans

1 can black beans

5 corn tortillas

1 can diced tomatoes

1 cup frozen corn

1 jalapeno, diced

1 onion, diced

2-15 oz cans vegetable broth

1 tsp cumin

1 tbsp minced garlic

1 tsp pepper

1 tsp cumin (if desired)

Cut tortillas into thin strips and place on a cookie sheet. Spray lightly with olive oil and bake in the oven at 350 degrees for 10 minutes. Combine corn, onion, kidney and black beans, jalapeno, and spices in a large pot. Cook on medium heat until onions are transparent. Add in diced tomatoes and vegetable broth. Cook until hot and serve immediately.

Mexican Rice

1-15 oz can black beans

1-15 oz can kidney beans

1 cup frozen corn

1 medium onion, diced

1 can green chilies

1 tsp cumin

1 tbsp chili powder

1 tbsp minced garlic

1 tsp onion powder

½ tsp cayenne pepper

1 tbsp olive oil

2 cups cooked brown rice

1 medium green bell pepper, diced

1 medium red bell pepper, diced

Sautee bell peppers, onions

Cook rice according to the instructions. In a medium skillet, sautee peppers, onions, and corn in oil. Add seasonings. Cook until vegetables are slightly soft. Add in cooked rice, green chilies, and rice. Stir well and then remove from heat. Serve while hot.

Tofu Scramble

1 container of firm tofu

1 small tomato, diced

1 bunch of green onions, diced

½ cup diced red and green bell pepper

1 small onion, diced

½ cup diced squash

½ cup frozen okra

½ cup diced zucchini

1 tsp fresh cilantro

1-2 tsp turmeric

Pepper to taste

Drain tofu. In medium skillet, combine all ingredients except the tofu. Scramble together until vegetables are slightly cooked but still firm. Add turmeric and pepper to taste.

Black Bean Salad

15 oz can black beans

1 cup frozen corn, thawed

1 jar salsa

Garlic to taste

Cumin to taste

Combine all ingredients in a bowl. Serve chilled over Roma lettuce with red

cabbage or wrap in a leaf of Roma lettuce.

Tofu Fried Rice

2 cups cooked brown rice

1 container extra firm tofu

1 large green bell pepper, diced

1 large red bell pepper, diced

1 can water chestnuts

15 oz can bean sprouts

1 large bag stir fry vegetables (no sauce)

1 tbsp soy sauce

1 tbsp olive oil

Salt and pepper to taste

Drain and cube tofu. Place a large skillet over high heat and add olive oil. Cook tofu in oil until it is slightly golden brown, adding salt and pepper to taste. Remove tofu from skillet and set aside. Cook vegetables and bean sprouts in oil until tender crisp. Add soy sauce and rice. Return tofu to the skillet. Serve meal immediately.

Sweet Potato Fries

1 large sweet potato

Olive oil spray

Preheat oven to 350 degrees. Wash sweet potato. Do not peel. Cut into small wedges and place on a cookie sheet in a single layer. Spray lightly with olive oil and place cookie sheet in oven. Cook for 15-30 minutes until sweet potatoes reach desired crispness. (Tip: If you line the cookie sheet with aluminum foil before adding sweet potatoes, you will get a crispier consistency. Sweet potatoes will not reach the same crispness that regular potato fries do.)

Apple Cinnamon Raisin Oatmeal

1 cup oats

Water (according to package instructions)

1 apple, peeled, cored, and diced

½ cup raisins

1 tsp honey

1 tsp cinnamon

Soy milk

1 tbsp milled flaxseed

Bring water to a boil. Add oats and apple. Cook until thick. Add raisins, cinnamon, and honey. Brown sugar can also be added for extra sweetness. Serve warm with soy milk and flaxseed.

Lard-Free, Vegetarian Refried Beans

1-15 oz bag pinto beans

Water to cover pot

2 tbsp minced garlic

1 jalapeno, diced

1 tbsp cumin

1 tsp pepper

1 tsp cayenne pepper

1 tsp onion powder

Sort and rinse beans according to package directions. Place all ingredients in a large pot or crock pot. Turn on low heat and simmer until beans are soft. (5-8 hours – add water to pot if necessary.) Once beans are soft, drain any additional water. Place beans in a food processor and blend until smooth. Use beans for any recipe you would normally use for refried beans.

Vegetarian Chili

1-15 oz can of black beans

1-15 oz can kidney beans

1-15 oz can pinto beans

1-15 oz can diced tomatoes

1-15 oz cans tomato sauce

1 cup frozen corn

1 medium onion, diced

1 green bell pepper, diced

1 red bell pepper, diced

1 cup diced carrot

1 tbsp chili powder

1 tsp onion powder

1 tsp cumin

1 tsp cayenne pepper

1 tbsp minced garlic

¼ cup nutritional yeast

1 tsp psyllium or flaxseed

1 tbsp olive oil

1 cup quinoa flakes

Sautee vegetables and kidney beans in oil and cook until onions are translucent. Add all other ingredients and cook until hot. (Tip: the longer you simmer chili, the spicier it will be.)

Tortilla Chips

10 corn tortillas

Olive oil spray

Preheat oven to 350 degrees. Cut tortillas into thin strips and place on cookie sheet. Spray lightly with olive oil. Cook until golden brown.

Healthy Nachos

1-15 oz can kidney or black beans

1 can olives, diced

1 medium onion, diced (green onions are fine as well)

1 small tomato, diced

1 tsp cumin

1 tsp garlic

1 tsp olive oil

Combine oil, kidney beans, and spices in a small skillet. Cook until warm. Pour over tortilla chips. Add tomato, onion, and olives. Veggie shreds can also be added in place of cheese.

<u>Trail Mix</u>

1 cup almonds

1 cup cashews

1 cup raisins

1 cup dried cherries

Place all items in a Tupperware container. Can be used for a snack in the middle of the day. You can also add other nuts or dried fruits or low fat pretzels.

Other vegetarian recipes can be found online from many different sources. Some of the most popular include

- PETA.org
- vegetariantimes.com
- allrecipes.com
- vegetarian.betterrecipes.com
- vegweb.com
- vrg.org
- cok.net
- chooseveg.com.

In addition to diet, we have created a natural cholesterol lowering supplement called CholestOFF. You can find it online at www.mycholesteroff.com.

The real point of this book, however, is to have you begin to question whether or not really have to lower your cholesterol in the first place. We also want you to question if medication is actually necessary to do so. We want you to break free of your health conditions that have been created by prescription statin drugs and start to regain control over your health.

Chapter 7:

Do You Have Neuropathy? Is There Healing?

Getting You on the Road to Recovery

Statin Drugs and Neuropathy

We have already discussed the symptoms of statin side effects. However, it is time that we revisit these symptoms. We want you to think seriously about your current health condition. We want you to consider, for just a moment, that there is healing for your painful ailments.

Neuropathy symptoms often depend on the specific nerves that are affected. The following symptoms are indicative of possible neuropathy.

- Tingling, numbness or pain
- Extreme sensitivity to touch
- Burning or freezing sensation
- Cramps and muscle spasms
- Prickling sensation
- Fatigue
- Paralysis
- Death

Some researchers estimate that 1 in 10 people who take statin drugs will experience a mild form of neuropathy where the symptoms may be pain or even a feeling of tiredness. Difficulty arising from a low chair, shortness of breath or difficulty walking can also occur.

In addition to neuropathy, there are other statin-induced side effects that you should be aware of:

- Memory loss
- Mood swings
- Kidney dysfunction
- Kidney failure
- Headache
- Dizziness
- Bloating/gas
- Nausea/vomiting
- Sleeplessness
- Flushed skin
- Diarrhea
- Rash
- Constipation
- Abdominal cramping or pain
- Muscle inflammation
- Rhabdomyolysis (is the breakdown of muscle fibers resulting in the release of muscle fiber contents (myoglobin) into the bloodstream. Some of these are harmful to the kidney and frequently result in kidney damage)
- Increased risk of heart attack

The list of side effects is long when it comes to statin drug use. However, neuropathy, muscle pain, and muscle weakness are three of the main side effects. Chiropractors are directly in line to visit the patients with these side effects. Chronic pain and pain not relieved by medical means drives millions of people to the chiropractic office.

Unknowingly, these chiropractors can face many clinical failures in cases that are not caused predominately from spinal malfunction, but by toxic neuropathy. Chiropractors need to be more aware of the damaging effects of statin drugs on their patients. They should also be aware of how failing to recognize statin-induced neuropathy can affect their own clinical outcomes, especially with the rising number of patients taking the drugs.

Peripheral Neuropathy Brings Pain and Weakness

"Peripheral" refers to nerves in the body's arms, legs, fingers, and toes. "Neuropathy" indicates damage to the nerves. This means that the messages sent out from the brain and back again are not working correctly. A disturbance of these motor nerves ends up leading to muscle weakness.

Other cases of peripheral neuropathy affect the nerves of the skin and sensory organs that tell the brain how the body feels. A defect in these sensory nerves leads to pain or numbness.

Leg, foot, or toe pain that includes burning, shocking, tingling, stabbing, or aching are all common neuropathy signs. This means that one or more of your sensory nerves to your legs and feet might be on the blink. If a motor nerve is involved, a person may experience difficulty lifting their foot off of the ground when walking. Peripheral neuropathy is one of the most common diseases, affecting more than 20 million Americans. Yet, few people have ever heard of it.

A poll of peripheral neuropathy patients found that the majority of people started having neuropathy symptoms during their prime adult years, with a shocking 82% experiencing the onset of the disease between the ages of 30 and 69. Polls also show that only a shocking 7% of all adults are aware of peripheral neuropathy. National Neuropathy Week, which starts May 17th, works to increase public awareness of neuropathy.

You probably know someone with the disease or, you may be suffering from the debilitating symptoms of peripheral neuropathy. Seeking diagnosis and getting treatment as soon as possible is extremely important. The longer the pain goes untreated, the more severe the condition can become. If left undiagnosed, peripheral neuropathy can lead to deterioration of the muscles and paralysis.

We all need throat muscles to swallow and chest muscles to breathe. What's more, the heart, the very thing that statins claim to protect, is a muscle. In extreme cases, severe neuropathy as a side effect to statin drug use, can lead to death.

Easing the Burden of Neuropathy Pain

Neuropathy means much more than just damaged nerves. It means damaged lives. Recent studies show that people with neuropathy have a much lower quality of life than people with chronic conditions like cancer, heart failure, Parkinson's disease, and stroke.

Finding a way to reduce the burden of neuropathic pain can be frustrating to patients because so few treatment options are offered. Sometimes, they are even told that "there is nothing that can help." However, recent advances in science have brought treatments that have been shown to reduce neuropathy pain and help restore nerve function.

We found that, in our office, when people figure out that neuropathy is the cause of their problem, they often do not know where to turn. This is exactly why we created The Neuropathy Treatment Centers of America – a patient advocacy group that is comprised of practitioners throughout the U.S, Canada, Mexico, Europe, and Africa who are pulling together to give information, treatment, and care to people suffering from peripheral neuropathy.

Since May of 2008, more than 400 doctors have enrolled in the training to become Certified Neuropathy Professionals and we are doing a pretty good job.

The ideal treatment for neuropathy is finding its cause and addressing that matter- whether it is from statin drug use or another source. Some other sources of neuropathy can include diabetes, deficiency of vitamin B-12, hereditary conditions, and a misguided attack on the nerves by the immune system. Quite often, a cause cannot be found. This type of neuropathy is referred to as "idiopathic."

Once the source of neuropathy is found, Low Level Light Therapy can be used with additional forms of treatment used by the NeuroTCA. These treatments have been shown in studies to provide more than 90% of patients with relief in just two weeks.

At a research conference in July of 2010 for NAALT (the North American Association of Laser Therapists), the members of the NeuroTCA and I were able to present some of our current research done. This is where my Master's Degree in Clinical Research really helped out. [Ref. http://www.naalt.org/2010-conference.htm]

Our research was used to determine the level of care and treatment satisfaction that patients with neuropathy experienced in my office. We also wanted to determine how our care compared to the care received by patients around the country with peripheral neuropathy. All participating treatment centers were members of the NeuroTCA.

In our study, we used a multi-modal care plan that included our patented Infrared Light Emitting Diode applications, ProneuroLight that was perceived as satisfactory for patients that have been diagnosed with peripheral neuropathy pain. We put our care to the test.

We started with an internet-based prospective survey in which practitioners from across the United States could enter the patient responses to their treatment at 5, 10, or 20 visits, answering the question, "Are you satisfied with your care?" We kept it pretty simple and straightforward.

Here is what we found:

- Seventy patients from 17 clinics responded between September and December of 2009.
- The average age of patients was 70 years old, 60% of which were female. (n=42)
- Identifiable risk factors for peripheral neuropathy pain were diabetes in 42.9% of patients (n=30), statin drug users in 27.1% of patients (n=19), and lumbar surgery in 8.6% of patients (n=6). Several patients had more than one risk factor and no definitive cause could be determined in 24.2% of patients (n=27).
- Most patients responded subjectively with a marked decrease in pain after only 5 to 10 treatments whereas others required several more treatments.
- After an average of 10.3 visits, 90% (n=63) were satisfied with the care. Only one patient, 1.43%, was not satisfied and six patients, 8.6%, were undecided. All seven of these patients had painful idiopathic neuropathy.

Table 1. Risk factors for peripheral neuropathy symptoms and treatment outcom

				Satisfaction
Risk Factor (n)	Age[a]	% female[a]	No. Visits[a]	Y/U/N
Diabetes (30)	65.2	88.7	9.3	30/0/0
Statins (19)	71.7	47.4	11.5	19/0/0
Lumbar surgery (6)	60.3	50.0	10.0	6/0/0
NN (27)	73.1	55.6	8.9	20/6/1
Total (82)	70	60.0	10.3	

a = mean; n= number of patients; NN = None noted; Y = yes, U = undecided, N = r

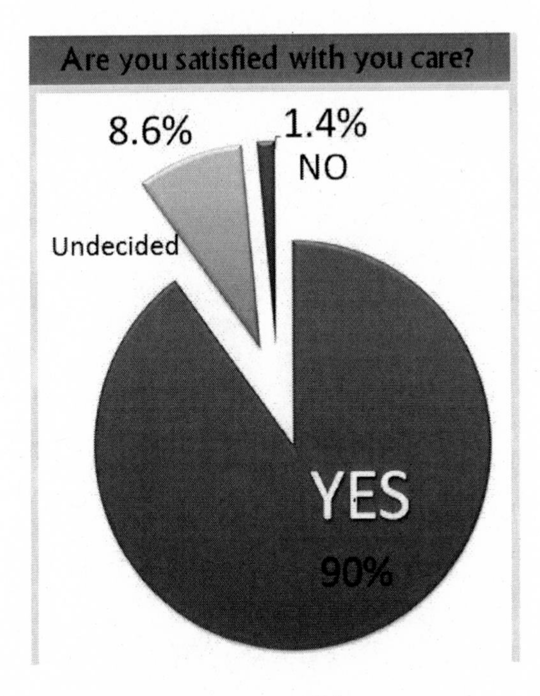

From our study, we can conclude that a large percentage of patients suffering from peripheral neuropathy pain appear to be satisfied with a multi-modal care plan that includes IR LED applications at the Neuropathy Treatment Centers of America.

The really amazing thing was that patients with neuropathy pain, as well as those with diabetes had a satisfaction rate of 100%. Only the patients with idiopathic or "unknown" causes of neuropathy were undecided or unsatisfied. These patients had a multitude of odds against them. Yet, even in this group, 74% of these difficult patients were satisfied.

It is clear from the studies that Low Level Light Therapy can be helpful in treating neuropathy from all sources, including statin drug use.

Chapter 8:

Putting it All Together - The Challenge

Now that you understand the truth about statin drugs and neuropathy, you may be wondering how to pull everything you have learned together. That's what this chapter is all about.

Hopefully, this book has given you the information you need to arm yourself against the onslaught of medical terrorism. If you do not stand up for yourself when it comes to your healthcare decisions, you will be the victim of the medical and pharmaceutical attacks on the American public.

My goal was to reveal the facts and the science of the cholesterol controversy that exists today and give you a way out of the storm. Remember that, if you choose to talk to your doctor about stopping your statin drugs, that you, unlike the hundreds of drug representatives that stream into your doctor's office every day, are not a sales person. These drug sales companies make money when you take statin drugs, the manufacturing companies make money, your local pharmacy makes money, and, truth be told, so does your doctor.

You are just Molly or Joe Citizen. What reason does your doctor have to listen to you? What you need to realize is that this is YOUR health, no one else's. Sometimes, as an individual, this means making an informed decision about your health, despite the advice of your doctor.

Of course, I can understand your initial concern. And, you are likely to get a concerned inquiry from your doctor, if and when, you finally decide to tell them that you are no longer taking statin drugs. You may be wondering, "What are my risks of stopping statins?" Your doctor is wondering, "What happens if he/she has a heart attack and sues me?"

There is no doubt that you have heard of all of the risks of stopping statin drugs – heart attack, stroke, etc. However, I hope by now that you realize, with the information you have been given, that you understand that the claims that are made by the drug companies have been spoon fed to your medical doctor, and to you. And truthfully, in some cases, these claims are exaggerated. And, as for your doctor, if the main concern of your doctor is to avoid a lawsuit, you must ask yourself, is this really in your best interest or in the best interest of your doctor.

Only you can decide for sure if it is best to talk to your doctor before stopping your statin drugs. However, you need to keep in mind that your doctor may try to convince you that it is not a good idea. If you buy into it, then you are likely to continue taking your statin drugs. This, of course, means that your condition will only get worse over time.

What we are attempting to do with this book is give you the real deal facts about cholesterol and some tools that will help you create a healthy cholesterol concept in both your mind and your body. We really don't advocate the lowest cholesterol levels possible because we have already seen the havoc that it can create in our bodies, our minds, and our nervous system.

What we really want to see is a transformed idea of what a good diet really is. We want to remove the crutch that society has placed upon the idea of taking a drug to "fix" your problem. Too many people assume that, since they are taking a statin drug that they can "cheat" and eat foods that are not good for their bodies. Instead, we want you to think about what you eat and why you eat it.

We aren't saying that you can't ever enjoy your favorite foods again but you need to learn what moderation is. You need to learn that an occasional "cheat," say, on holidays can be worked in and managed, as long as you are working hard at keeping your diet healthy and exercise the rest of the time.

Of course this is hard. We understand this. This is why you have this book. We want you to use the knowledge you have gained, implement some common sense, and learn how to manage your health properly. The sooner you take your health into your own hands, the greater your chances of neuropathy recovery may be.

If, by chance, you don't have peripheral neuropathy currently, know that this is the best time to stop taking your statin drugs. For many people, it is only a matter of time before neuropathy starts. If, however, you do have neuropathy, know that there is hope.

Stopping your statin drugs is the first step. This is, very likely, the cause of your neuropathy pain if you are currently taking them. Your next step is to start eating a healthy diet and exercising regularly. Even if your cholesterol levels don't need managed, this is important to maintaining a healthy lifestyle.

If you are reading this book and you suffer from neuropathy, you may also be wondering, in addition to changing your diet and stopping your statins, if there is anything you can do to get rid of your neuropathy pain. You may wonder if there are specific treatments that are necessary to helping neuropathy pain. I suggest that you consider an Advanced NeuroFoot Analysis with a certified neuropathy professional at one of our Neuro TCA offices. There are 450 clinics across the country and home care is also available.

Is there a cure for peripheral neuropathy? In some cases, we can say YES. In some cases, this chronic disease needs to be managed, not cured. Of course, before these questions can only be truly answered after a proper evaluation.

If you would like to know more about the treatment offered through the Neuro TCA, we encourage you to contact us. The treatment is non-invasive, painless and no drugs are involved. Treatment is offered for all neuropathy suffers, regardless of what the cause or source. You can reach the NeuroTCA at 877-296-7799. When you call, ask for an Advanced NeuroFoot Analysis. You can also find us on the web at www.neurotca.com.

References

1. Jacobs MB: "HMG-CoA reductase inhibitor therapy and peripheral neuropathy."

Ann Intern Med. 1994 Jun 1;120(11):970.

2. Formaglio M, Vial C: "Statin induced neuropathy: myth or reality?" Rev Neurol

(Paris). 2006 Dec;162(12):1286-9.

3. De Langen JJ, van Puijenbroek EP: "HMG-CoA-reductase inhibitors and neuropathy: reports to the Netherlands Pharmacovigilance Centre." Neth J Med.

2006 Oct;64(9):334-8.

4. Law M, Rudnicka AR: "Statin safety: a systematic review." Am J Cardiol. 2006

Apr 17;97(8A):52C-60C.

5.http://www.timesonline.co.uk/tol/life_and_style/health/article2127605.ece

6. Golomb BA, McGraw JJ, Evans MA, Dimsdale JE: "Physician response to patient reports of adverse drug effects: implications for patient-targeted adverse effect surveillance." Drug Saf. 2007;30(8):669-75.

SOURCES

1. U.S. Food and Drug Administration website, May 2010, [http://www.fda.gov/ForConsumers/ByAudience/ForWomen/ucm118595.htm#hmg.

2 Golomb BA, McGraw JJ, Evans MA, Dimsdale JE., Department of Medicine, University of California San Diego, La Jolla, CA 92093-0995, USA. bgolomb@ucsd.e (Medline)

3. First comprehensive paper on statins' adverse effects released, Published: Tuesday, January 27, 2009 - 10:15 in Health & Medicine

4. http://docnews.diabetesjournals.org/content/4/12/1.1.full

5. [http://www.washingtonpost.com/wp-dyn/content/article/2007/08/24/AR2007082401714.html].

6. PMID: 15934847 [PubMed - indexed for MEDLINE

7. http://www.ncbi.nlm.nih.gov/pubmed/12011277

8. JAMA. 2006 Jan 4;295(1):74-80

9.JACC .... http://www.drbriffa.com/blog/2008/08/22/low-cholesterol-levels-linked-with-increased-risk-of-cancer-so-is-cholesterol-reduction-safe/ or (?) Alawi A, et al. Statins, Low-Density Lipoprotein Cholesterol, and Risk of Cancer. Journal of the American College of Cardiologists. Published on-line 20th August 2008.http://content.onlinejacc.org/cgi/content/short/j.jacc.2008.06.037v1

10. [source: Alawi A, et al. Statins, Low-Density Lipoprotein Cholesterol, and Risk of Cancer.

Journal of the American College of Cardiologists. Published on-line 20th August 2008.]

11. Am J Cardiovasc Drugs. 2008;8(6):373-418. doi: 10.2165/0129784-200808060-00004 / PMID: 19159124 [PubMed - indexed for MEDLINE]PMCID: PMC2849981Free PMC Article http://www.ncbi.nlm.nih.gov/pmc/articles/PMC2849981/ // http://www.ncbi.nlm.nih.gov/pmc/articles/PMC2849981/table/T3/

12. [http://www.time.com/time/magazine/article/0,9171,921647-9,00.html].

13.[http://www.pfizer.com/investors/financial_reports/annual_reports/2009/iob-thenumbers.jsp].

14. http://www.jigsawhealth.com/resources/coenzyme-q10

15.http://seattletimes.nwsource.com/html/nationworld/2008371545_heart10.html

16.[http://www.cspinet.org/integrity/press/200409231.html].

17. Press release on Center for Science and the Public Interest Website. Get link from site.

18.http://online.wsj.com/article/SB10001424052748703685404575307021820487134.html

19. PMID: 7663036 [PubMed - indexed for MEDLINE]

20. Indian Heart Journal 2002; 54(5). "A Comparative Evaluation of Ispaghula Husk (Dietary Fiber) and Simvastatin in Hyperlipidemic Indian Patients"

21. Arch Intern Med. 2005;165:1161-1166

22. THE LEAN PLATE April 03, 2006|Sally Squires, Special to The Times (find link)

23. www.vegparadise.com

24. find link

25. Soy Rich diet, http://www.soyfoodsillinois.uiuc.edu/PortfolioPlanSoy.pdf

26. http://www.naalt.org/

27. http://www.whitakerwellness.com/health-concerns/what-you-dont-know-about-statins-can-hurt-you/